AF316742

The Space Roc Band Presents:

Rockin' Around the Sun

By Goddess Chappelle

Copyright © 2022 by Goddess Chappelle

All rights reserved.
No part of this publication may be reproduced in whole or in part, or stored in a retrieval system, or transmitted in any form or by any means, electronic, mechanical, photocopying, recording, or otherwise, without written permission of the author.

☀️Pour wisdom into your mind to fuel your superpowers.
-Goddess Chappelle

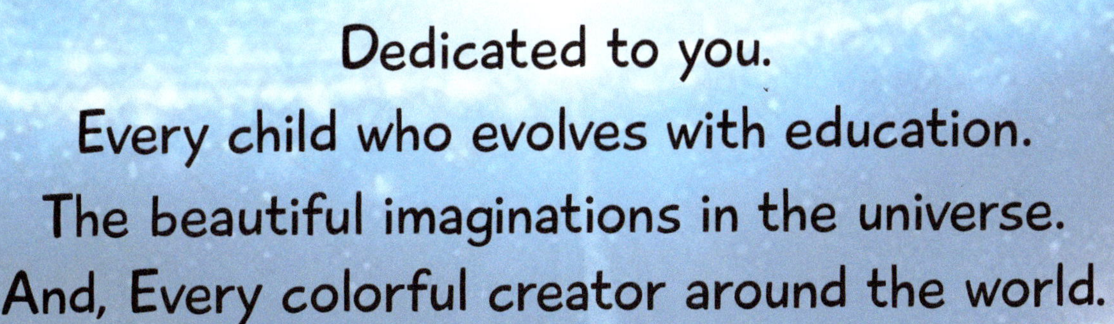

Dedicated to you.
Every child who evolves with education.
The beautiful imaginations in the universe.
And, Every colorful creator around the world.

Build yourself with knowledge and become a hero.

-Goddess Chappelle

Introducing The Space Roc Band

Benji Bari

Kat Mae

Sam Roc

Yaz Giant

Rhodi Zo

First is the Sun
It weighs more than a ton
If you stare, you will get stunned

So, grab your drum and move your bum
Rockin' around the Sun with fun

Next you will see that it's Mercury.
So, count to three
And make a beat.

Turn up the heat
And move your feet
Rockin' around the sun is neat.

This is genius, we just hit Venus
We dance and rock all in arenas.

Man, we are the coolest genus
Space Roc Band will always teach us
Rockin' around the sun is sweetness.

We hit a swerve on down to Earth
Space Roc Band is on our shirts.

We sing and dance all on the turf
We love Space Rock is what we spurt.
Rockin' around the sun with a quirt.

Learning space like superstars
We follow stars on down to Mars
Playing tunes on our guitars.

Hoping that we will go far
Jamming in our Rocket car
Rockin' around the sun with bars.

Learning space is so much cooler
Flying all the way to Jupiter

Planets make our noggins cooler
We are now the Space Roc Rulers
Rockin' around the sun like troopers

Bop and rock on down to Saturn
Glowing up just like Green Lanterns.

Hit the worm and do a turn
Our minds will grow like little ferns
Rockin' around the sun to learn.

We groove on down to Uranus
Our chromed out rocket car is stainless.

Techno neon color changes.
Singing rock music makes us famous
Rockin' around the sun in ranges.

Our Rocket Car goes vroom
We zoom on over to Neptune
All dressed up in space costumes.

We aim high just like balloons
Watch us hit our favorite moves
Rockin' around the sun with tunes.

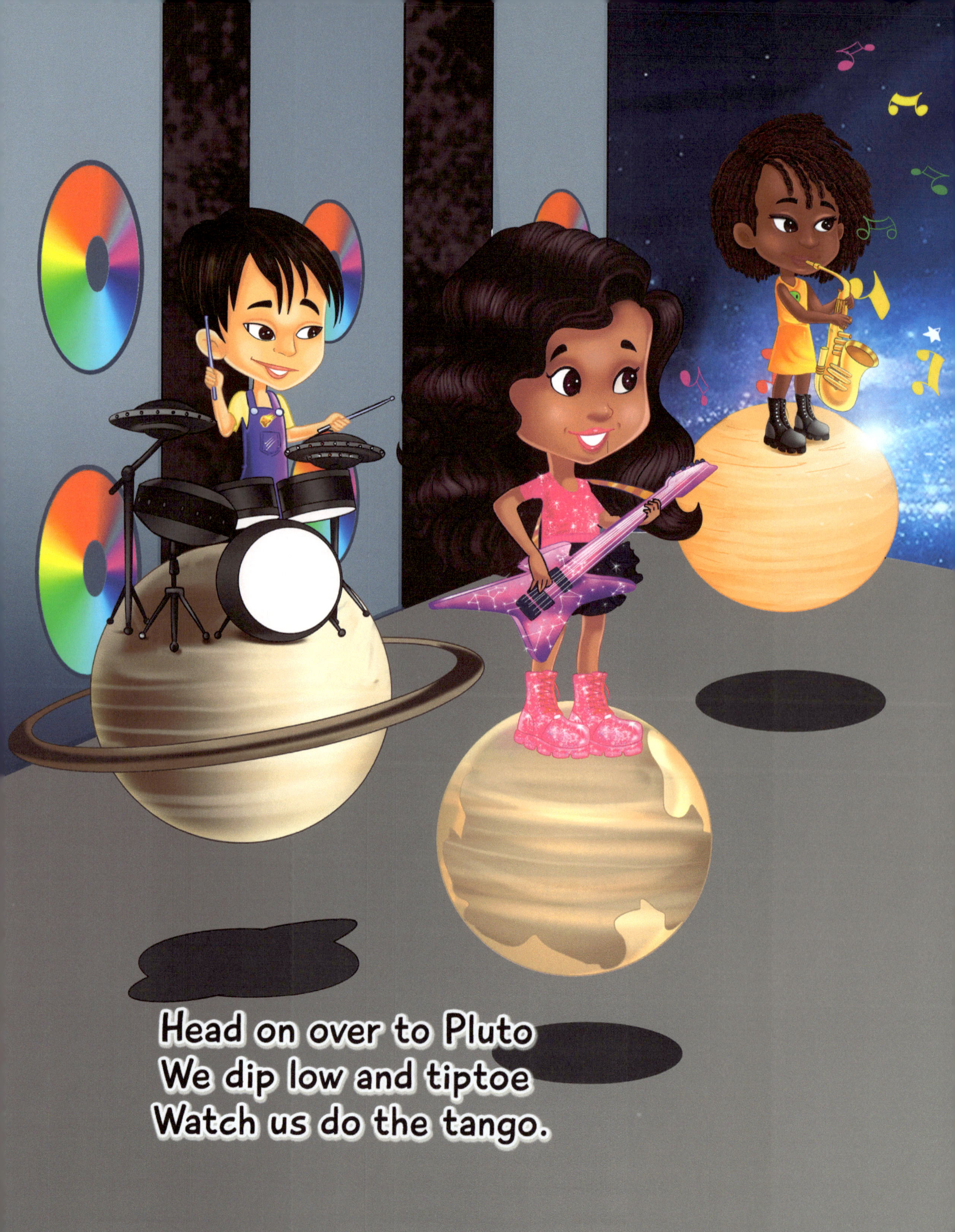

Head on over to Pluto
We dip low and tiptoe
Watch us do the tango.

We drive faster than Kangos
Rockin' around the sun like angels.

Did You Know?

1. One million earths could fit inside the sun.
2. Sirius is the brightest star in the night sky.
3. The universe is observed to be 13.8 billion years old.
4. Space is completely silent.
5. There is floating water in space.
6. On the planet Mercury, days are longer than years on Earth.
7. Pluto is now classified as a dwarf planet.
8. Pluto was named by an 11 year old girl.

Reach for light in the universe and you can be a star.

-Goddess Chappelle

About the author

Goddess Chappelle is a Mental Health Technician and CEO of Characters in Color LLC. Goddess is an advocate for confidence in education and, is committed to helping children build their knowledge and imagination.

Ignite your brain power and S.T.E.A.M through the universe.
-Goddess Chappelle

WHAT IS YOUR FAVORITE PLANET AND WHY?

DRAW YOUR OWN SOLAR SYSTEM.

www.ingramcontent.com/pod-product-compliance
Lightning Source LLC
Chambersburg PA
CBHW042322140726
48196CB00015B/705